How to design and develop business systems

CASE STUDIES

Steve Eckols

Mike Murach & Associates, Inc.

4222 West Alamos, Suite 101
Fresno, California 93711
(209) 275-3335

Development team

General editor:	Mike Murach
Technical editor:	Anne Prince
Copy editor:	Carrie Gwynne
Production director:	Steve Ehlers
Artist:	Ed Gallock

Printed in the United States of America.

20 19 18 17 16 15 14 13 12 11 10 9 8 7 6 5 4 3 2 1

Library of Congress Catalog Card Number: 83-62380

ISBN: 0-911625-18-6

Contents

Introduction

This workbook contains four case studies for you to use with *How to Design and Develop Business Systems*. Three of the case studies are examples of original system development projects; one is a modification to the brokerage system the text presents. In each case study, you'll use the techniques the text presents to analyze an existing system and to design a new one (or part of one).

Each case study includes a narrative that presents the information you need to solve the problems it presents. Some include samples of documents that are important to the application. Of course, on an actual system development project, you'd be responsible for gathering this information.

How the case studies are organized

Each of the three case studies that represents a new svstem is organized the same way. The material is presented in the same sequence you'd probably encounter it in an actual system development situation and in the same sequence as the system development phases I present in the text. Consequently, you need to work each case study from the beginning. The later exercises in each case study use the products of the earlier exercises.

To do the last of the four case studies, you need to be familiar with the brokerage system example in the text and all of the methods the text presents. As a result, you shouldn't try it until you've completed the text.

In some instances, your instructor may want you to do later exercises from a case study without doing the earlier exercises first. When that's the case, your instructor will have to supply you with the solutions to the earlier exercises.

Each of the case studies has seven sections, and each section has one or more exercises. The sections correspond to chapters 3 through 9 of the text. Since the first two text chapters are introductory, I didn't give you exercises for them.

Section 1: Analysis DFD Chapter 3 of the text teaches you how to develop an analysis data flow diagram. So, your first exercises in each of the case studies will be related to drawing an analysis DFD.

Section 2: Model DFD After you've drawn the analysis DFD, you'll learn what the requirements are for the case study's new system. Based on those requirements, you'll list the changes in functional requirements for the new system and develop a model data flow diagram for it. Chapter 4 of the text shows you how to do this.

Section 3: Data dictionary entries After you've identified the critical functions of the new system, as well as their interrelationships, you need to begin to determine what data elements the new system requires. Chapter 5 explains how to do this, and the case studies include exercises in data dictionary notation to give you practice using this skill.

Section 4: System structure chart When you design a system, you need to be sure you identify all of the system's functions and that you organize them in the most logical, useful way. Chapter 6 explains how to do this using the system structure chart. Each case study includes exercises that have you develop system structure charts and modify them to meet user needs.

Section 5: Database design When you've determined what all of the system functions are, you're in a position to decide how the system's database should be organized. In this case, much depends on the hardware configuration that will be used for the application. Each case study includes exercises to allow you to use the techniques of database design presented in chapter 7, including the preparation of data access diagrams and data hierarchy diagrams.

Section 6: Program specifications For each case study, I've selected an application program for you to specify. This includes designing reports and screens, as well as preparing a program overview. This material is correlated with chapter 8 of the text.

Section 7: Program design Finally, for each program you specify in the exercises for chapter 8, you may be called upon to prepare a program design using the techniques presented in chapter 9.

How you should approach the case studies

I designed these case studies to give you an opportunity to use the techniques the book presents and to do some creative thinking about systems problems. The point of the text is not to teach you how to design a computer system for a brokerage firm, and the point of these case studies is not to teach you how to design a system for a workers' compensation adjuster, a video cassette rental shop, or a vending machine business. The point of this course is to help you develop your analytical skills within the context of actual applications.

These case studies will take a lot of thought and effort on your part. But I'm convinced that if you invest the time they require, you'll enhance your system development skills. Not only will you be better able to handle a system development task more efficiently, but you'll also enjoy it more.

Steve Eckols
Fresno, California
January, 1984

Case Study 1

Workers' compensation system

An insurance partnership administers a workers' compensation plan for a large company. The partners want to install a computerized system to improve their efficiency and their service to the client company.

Background

If an employee is injured on the job, his employer is responsible for paying all medical costs related to the injury. In addition, the employer must make weekly payments to the employee during the time the employee can't work to help compensate for lost wages. In particularly serious cases, the employer must compensate the employee for permanent disabilities that result from the injury.

To protect themselves against potentially large workers' compensation claims, many companies take out an insurance policy. However, workers' compensation insurance policies have become increasingly expensive in recent years. As a result, more and more companies, especially those with low "exposures" to workers' compensation claims, are choosing to be "self-insured." That is, they assume the risk of paying workers' compensation claims themselves.

In the state in which the insurance partnership operates, a self-insured firm must employ a licensed outside adjuster to administer the self-insured plan. That's the service the insurance partnership in this case study offers. The administration of the plan involves (1) investigating claims, (2) issuing checks to pay claims from a reserve fund, and (3) preparing required government and management reports.

Section 1: Analysis DFD

Under the existing manual system, a file is maintained for each claim the partnership processes. The administration of the client's self-insured plan revolves around the files. In fact, the employees of the partnership refer to their jobs as "working the files."

When an adjuster receives a new claim from an injured worker (or from one of the client firm's personnel administrators), he creates a new file for the claim. He assigns a unique number to the claim and records the basic information summarized in figure 1-1.

Data items recorded by the adjuster when he opens a new claim:

claim number
claim type (medical only or medical with disability compensation
claim classification (regular, life pension, fatality)

claimant's name (last, first, middle initial)
claimant's social security number
claimant's sex
claimant's date of birth

date claim opened

injury description
injury codes (attachment)
injury date

department or group in which employee works

subrogation possibility (will the adjuster try to charge someone else with liability for the injury and, as a result, escape liability for paying?)
appeal filed to state insurance commission?
medical reserve
disability compensation reserve

Figure 1-1 (Part 1 of 2) Information recorded when a new claim is opened

One of the most important parts of opening a new claim is assigning reserve amounts to it. A single claim can have two reserve amounts specified: medical and disability compensation. A claim always has a medical reserve. Claims that may involve a disability are also assigned a disability compensation reserve. The amount of money reserved depends on the adjuster's judgement of how much the claim is likely to cost.

The reserve represents an amount of money the insured company must set aside to pay expenses incurred as a result of the injured employee's claim. The self-insured company must maintain a bank account that contains enough money to cover all current claim liabilities. (Current liabilities are total reserves minus total payments on active claims.)

As the adjuster works on a claim, he authorizes payments from that reserve bank account. (A payment is issued when a medical bill is received or a disability compensation payment is due.) The adjuster requests a check by writing a note on a piece of scratch paper that's attached to the outside of the claim file. Then, he passes the file to a clerk. The clerk writes the check, then returns the file with the check to the adjuster.

The adjuster reviews the check and, if it's correct, mails it to the payee and records the payment in the claim file.

The claim file contains data for each check issued on that claim. Payments are classified into five categories: medical, disability compensation, legal fees, investigation fees, and other fees. A single payment may be broken down into two or more of these categories. For example, a claimant may be issued a check for $320, $220 of which is for disability compensation and $100 of which is reimbursement for medical expenses. It's important that the distribution of each payment into these five categories be recorded because state regulations require that payments be reported by category. To calculate the current liability on a claim, medical payments are deducted from the medical reserve; the other four categories of payments are deducted from the disability compensation reserve.

When all anticipated medical and compensation expenses for a claim have been met, the adjuster closes the claim by making a notation in the file. (The adjuster knows when all payments have been made because he's in continuous contact with the claimant and the insured firm.) Of course, the claim file is retained, but the claim is no longer active. Any funds remaining on reserve for the claim are released and can be allocated to new claims. Sometimes, the claim file is reopened if the claimant has later problems resulting from his original on-the-job injury.

Twice each year, the self-insured firm must submit a two-part claim report to the state insurance commission. The first part of the report lists this data for each open claim, in order by the claimant's last name:

claimant data: name, age, sex, social security number, date of birth
injury data: description and date
original reserves
total payments made by category

The second part of the report consists of summary information that shows the total of reserves and payments made on open claims.

The self-insured company itself requires a similar report for management purposes. It contains all of the information on the state-mandated claim report, plus detail data on all checks issued during the reporting period.

KEY TO CODE

TYPE

A. Striking against
(Such as bumping into, stepping on, kicking, being pushed, or thrown against objects)

B. Struck by
(Falling, flying, sliding, or moving objects)

C. Caught in, under or between
(Pinched or crushed between moving objects, a moving and stationary object, moving parts of an object)

D. Fall on same level

E. Fall to different level or fall from elevation

F. Bodily reaction
(Strains, sprains, ruptures, etc. from an unnatural position or induced involuntary motions)

G. Contact with temperature extremes
(Consists of burn, heat exhaustion, freezing, frostbite, exposure)

H. Contact with radiations, caustics, toxic and noxious substances
(Causing poisoning, drowning, chemical burns, allergic reactions, etc.)

I. Overexertion
(Lifting, pulling, pushing, wielding, throwing, or handling of objects)

J. Contact with electric current

K. Rubbed or abraded
(Produced by pressure, vibration, friction, or foreign matter in eyes)

L. Vehicle accidents
(Motor, railway, water, aircraft, etc.)

M. Unclassified, insufficient data

N. Bitten

O. Altercation

P. Continuous trauma

Q. Occupational disease

PART OF BODY

1. Head
2. Eye
3. Shoulder
4. Arm/Elbow
5. Hand and/or Wrist
6. Finger
7. Leg/Knee
8. Foot/Ankle
9. Toe
10. Ribs and/or Chest
11. Back
12. Abdomen
13. Multiple Parts
(Applies when more than one major body part has been affected, such as an arm and a leg)
14. Body Parts
(Not elsewhere classified)
15. Respiratory System
(Lungs, etc.)
16. Circulatory System
(Heart, blood, arteries, veins, etc.)
17. Musculo-skeletal System
(Bones, joints, tendons, muscles, etc.)
18. Digestive System
19. Excretory System
(Kidneys, bladder, intestines, etc.)
20. Unclassified
(Not filed through our office or insufficient information to identify part affected)
21. Ear
22. Teeth
23. Hip
24. Facial Area
25. Uro-Genital
26. Cervical
27. Thoracic
28. Lumbo-Sacral
29. Needle Puncture

NATURE

A. Amputation

B. Asphyxiation, Strangulation, Drowning

C. Burn or Scald (Heat or Chemical)

D. Inflammation or Irritation of Joints, Tendons, or Muscles—includes bursitis, synovitis, tenosynovitis, etc. (Does not include strains, sprains, or dislocation of muscles or tendons)

E. Contusions

F. Conjunctivitis

G. Concussion, Shock, Prostration

H. Dermatitis-Rash, skin or tissue inflammation, including boils, etc. (Does not include inflammation or irritation resulting from friction or impact)

I. Fractures

J. Cut, Laceration, Puncture—open wound and infections

K. Poisoning, systemic

L. Strains, Sprains or Dislocations

M. Pneumoconiosis—including anthracosis, berylliosis, asbestosis and silicosis

N. Ulcerations

O. Varicosities

P. Multiple Injuries

Q. Electric Shock, Electrocution

R. Other Nature or Unclassified

S. Death

T. Foreign Body other than eye

U. Hernia

V. Occupational Disease—Chemical

W. Occupational Disease—Non Chemical

Figure 1-1 (Part 2 of 2) Information recorded when a new claim is opened

To prepare either of these claim reports, a clerk reads through all of the claim files, accumulates the appropriate reserve and payment figures, and types the report. Both of these reports are dozens of pages long.

Each month, a clerk also prepares a reserve requirements report. It contains information drawn from the claim files and specifies the reserve amounts that need to be added to the claim fund bank account. That amount is based on the difference between reserve requirements of new claims and reserve funds released when claims are closed. If the former is greater than the latter, the client firm must deposit the difference.

Exercises

1-1-1 Identify the inputs to and the outputs from the existing workers' compensation system.

1-1-2 Draw the context DFD for the existing workers' compensation system.

1-1-3 Identify the data stores of the workers' compensation system.

1-1-4 Draw an analysis DFD for the existing workers' compensation system.

Section 2: Model DFD

The insurance firm wants to cut the clerical expense required to administer its client's workers' compensation plan and administer it more efficiently and accurately. The partners in the firm are convinced that installing a computerized workers' compensation system will help them meet these objectives.

A major goal of the new system is to eliminate the process of issuing checks manually. Under the present system, an adjuster writes out a request for a check, and a clerk issues it. Under the new system, the adjuster will request a check at a computer terminal, and it will be printed immediately.

Another goal of the new system is to prepare required government and client reports by computer. This will be a major time-saver and will eliminate many clerical errors.

And finally, for the adjuster's day-to-day information needs, the paper claim files will be replaced by inquiry programs at computer terminals. Information on claim activity should be available for display on the adjuster's terminal. And the adjuster should be able to access a claim either by number or by the claimant's last name.

Exercises

1-2-1 Identify the changes in functional requirements for the new workers' compensation system.

1-2-2 Identify the data flows you can omit from your first model DFD for the new workers' compensation system.

1-2-3 Create a model DFD for the new workers' compensation system.

Section 3: Data dictionary entries

1-3-1 Using data dictionary notation, create preliminary contents lists for the data stores on the model DFD for the new workers' compensation system.

Section 4: System structure chart

1-4-1 Working from the model DFD, create a preliminary system structure chart for the new workers' compensation reporting system.

1-4-2 Expand the processes on the preliminary system structure chart. To do so, use a table similar to the one in figure 6-3 of the text (page 127). Then, redraw the system structure chart to include the new functions you've identified.

1-4-3 Add other required functions to the system structure chart that aren't part of the model DFD. If necessary, expand them to show their important component functions. Be sure to group related functions to keep the span of control of the top-level system structure chart modules reasonable. Name and number each system structure chart module.

1-4-4 Redraw the system structure chart for the workers' compensation system to include a backup function.

Section 5: Database design

1-5-1 In this application, data for all checks issued on any particular claim must be available for reporting and inquiry. Should check data be maintained in the claims data store or in a separate data store? Why or why not?

1-5-2 Create a new contents list for a checks data store. Then, modify the contents lists you developed in exercise 1-3-1 to reflect the addition of the checks data store.

1-5-3 Identify the access paths to the data stores of the workers' compensation system. Draw a data access diagram and a data hierarchy diagram to show the relationships among the data stores and the access paths of this system.

1-5-4 What factors would you consider if a co-worker suggested including summary payment fields in the claim record as well as the detailed information that's included in the separate check records?

1-5-5 Suppose the insurance firm wants to provide similar administrative services for other companies. Discuss how this business decision would affect your system design.

Section 6: Program specifications

1-6-1 Which modules on the system structure chart for the new workers' compensation system are likely candidates for combination?

1-6-2 Prepare complete program specifications for the system structure chart module the adjuster will use to void incorrect checks.

Be sure to include these components:

program overview
screen layouts
report layout for the audit trail document the program should create

Section 7: Program design

1-7-1 Using the program specifications from exercise 1-6-2, create a program structure chart for the void-claim-checks program.

Case Study 2

Video cassette rental system

The owner of a small video cassette rental shop is planning to install a computer system to keep track of his customers and inventory. It's your job to analyze the existing operation and required changes, and then create a design for the new system.

Section 1: Analysis DFD

Basically, the operation of the video cassette rental shop involves maintaining records on (1) customers, who are required to buy memberships, (2) video cassettes that make up the shop's inventory, and (3) rental and return transactions.

Membership A customer must be a member of the shop's "movie club" before he can rent video cassettes. To become a member, a customer completes an application like the one in figure 2-1 and pays a membership fee of $50.00. The information the customer supplies on the application is basic identification: name, address, and telephone numbers. In addition, the customer must provide some sort of security for the movies he'll be renting. That security can be either a blank check, which remains attached to the application, or a credit card number and expiration date. If the customer fails to return a movie he rents, his check will be filled in and deposited, or his credit card account will be charged.

Once the application is complete and has been approved, the customer is assigned a membership number, and the clerk types a membership card for him. Then, the clerk puts the new customer's membership fee in the cash register and files the application by the member's last name.

Inventory Currently, the shop has approximately 500 movie titles on video cassette available for rental and a total of about 650 cassettes (some popular titles are duplicated in the collection). The owner decides to order tapes based on announcements of new releases from distributors and on requests from customers. Because volume is relatively small, the owner remembers the orders he has made and doesn't need to keep written records of them.

When a new movie is received from a distributor, the owner adds it to the collection. To prepare a movie for the collection, he assigns it a unique identification number and puts the cassette in a protective plastic case. Then, he completes a 3 x 5 card for the movie, like the one in figure 2-2.

MOVIE CLUB MEMBERSHIP APPLICATION

Member No: ____________ Date: ____________

Name: ____________________

Address: ____________________

City: ____________________ State: _____ Zip code: ________

Phone: (h) ____________

(w) ____________

Security: _____ VISA _____ MasterCard _____ American Express

Number: ____________________

Expiration Date: ________________

_____ Personal Check

I understand that I may retain a rented movie for no longer than ten (10) days. If I keep any rented movie for more than 10 days, I will forfeit my membership in the "movie club" and will be required to pay for the movies I have on rental with the check or credit card charge authorization included with this application.

Signed ____________________ Date ____________

Figure 2-1 Movie club application

ID No: ____________________ Purchase Date: ____________

Title: ____________________

Format: VHS_____ Beta_____ Purchase Price: ____________

Rental fee: ____________

Notes:

Figure 2-2 Sample cassette record card

These cards are filed by cassette identification number and specify the film's title, tape format, purchase date and price, and daily rental fee.

Next, the owner prepares three copies of an identification label. The label specifies the title of the film, the cassette's identification number, tape format (VHS or Beta), and the film's daily rental fee. Figure 2-3 is a sample of the identification label.

One copy of the label is attached to the plastic case that contains the cassette, and the second is attached to the cassette itself. The tapes are stored behind the rental counter to prevent theft. The inventory of movies is separated by tape format (VHS or Beta).

The third copy of the label is attached to the tape's original carton. Usually, these cartons are printed with flashy graphics and a brief description of the movie each contains. The empty cartons are stored on shelves in the main section of the shop so customers can select the films they want to rent. They're also separated by tape format.

Rental and return transactions To rent a movie, a member takes the cartons for the cassettes he wants from the shelves in the middle of the shop and completes a three-part rental form like the one in figure 2-4. The rental form includes spaces for the member's name and membership number, plus identifying information from the carton labels for up to four movies. The shop doesn't allow a member to rent more than four movies at a time.

When the member has completed the rental form, he presents it along with the cartons for the selected movies and his membership card to a clerk. The clerk picks the requested cassettes from the shelves behind the counter and replaces them with the cartons for those movies. That way, the cartons that are on the shelves in the middle of the shop always represent available movies.

Then, the clerk has the member sign the form, and the rental transaction is complete. The customer leaves with one copy of the rental form and the cassettes. The member doesn't pay rental charges in advance because he may keep the movies for one or several days, so the exact rental charge won't be known until he returns the cassettes.

The clerk keeps two copies of the form, as well as the customer's membership card. (It's required that customers leave their cards at the shop when they rent movies.) The

TITLE		
FORMAT V B	ID NO.	DAILY RENTAL

Figure 2-3 Sample movie cassette and carton label

VIDEO CASSETTE RENTAL

NO. 008581

Member No: ____________________ Date: __________

Name: ____________________
Address: ____________________
City: ____________________ State:______ Zip code:______
Phone: (h) ______________ (w) ______________

Title	Format (V/B)	ID No.	Daily rental fee	Days rented	Total rental fee

Rental due: ______

Sales tax: ______

Total due: ______

I understand that I may retain a rented movie for no more than ten (10) days. If I keep any rented movie for more than 10 days, I may be held liable to purchase it at its current retail price and may be required to forfeit my "movie club" membership.

Signed ______________________________ Date ______________

Figure 2-4 A sample video cassette rental form

clerk files the copies of the rental form along with the customer's membership card in alphabetical order by the customer's last name.

When the customer returns the movies, the clerk searches through the file of current rental forms to find the one for the movies being returned and the customer's identification card.

Then, the clerk calculates how much the customer owes for the rentals on the two copies of the rental form that were retained in the shop. Rental charges are based on daily rental fees for each cassette and the number of days the member kept them. When the customer pays, the clerk rings up the rental on the cash register. Then, the clerk gives the customer one copy of the completed rental form as a receipt for the returned movies and files the other by serial number.

When he has free time, the clerk returns the cassettes to their proper places on the shelves behind the counter and replaces the cartons for those films on the shelves in the middle of the store. That way, other customers will know those movies are again available for rental.

Once each week, a clerk searches through the file of current forms for rentals that are overdue. It's the shop's policy that a customer shouldn't keep a movie for more than 10 days. If a customer does keep a movie for more than 10 days, the clerk will try to get in touch with him to have him return it. The clerk makes a note on a slip of paper and clips it to the rental form each time he's in touch with a customer who has overdue movies.

If a clerk repeatedly contacts a customer, but the rented films still aren't returned, the owner of the shop will revoke that customer's membership and use the security the customer left with his membership application to pay for the overdue movies. Either the customer's check is filled in and deposited, or his credit card account is charged. This payment is rung up on the cash register like a normal rental. Then, a notation is made on the customer's membership application that his membership has been revoked, and his membership card is attached to the original application.

Each day, the owner prepares a bank deposit of the money accumulated in the cash register. The deposit includes cash, checks, credit card charge slips, and a bank deposit slip.

Exercises

2-1-1 Identify the inputs to and the outputs from the existing video cassette rental system.

2-1-2 Draw the context DFD for the existing video cassette rental system.

2-1-3 Identify the data stores of the video cassette rental system.

2-1-4 Draw an analysis DFD for the existing video cassette rental system.

Section 2: Model DFD

One of the major problems of the existing system is that it involves handling many pieces of paper. Too often, misplaced slips of paper cause expensive and embarrassing errors. Under the new system, all record keeping for basic membership, inventory, and rental and return transactions is to be automated, and paper records are to be minimized.

Membership Enrolling a new member should operate much as it does under the existing system. A new member will still fill out an application and offer some form of security for the movies he'll rent, and the clerk will still type a membership ID card for him. In addition, however, a computer record for the new member will be created that will be used to streamline the rental and return process.

The membership record maintained for each customer must contain information that will allow data to be displayed on a demand inquiry basis for a specific customer or on a batch reporting basis for all customers. The following data will be kept for each customer:

number of movies rented (cumulative)
number of movies rented (this year-to-date)
rental sales amount (cumulative)
rental sales amount (this year-to-date)

Also, the system must be able to display the current status of any member. In other words, the system must be able to display all of the movies a particular customer currently has on rental.

Inventory When a new movie is added to the collection, the owner wants the computer to prepare the identifying labels for it. Also, the system should keep a record for each cassette that will be used in rental and return transactions. It should also contain cumulative statistics on the number of times a particular cassette is rented. The rental status of any particular cassette must be available for display: is the cassette in the shop or is it checked out, and if it's checked out, who has it?

Remember that the shop may stock more than one copy of a specific movie. As a result, the owner of the firm wants information not only on specific cassettes, but also for specific titles. Summary statistics on each title must be accessible to help the owner keep track of how popular certain films are and to decide which ones to stock in multiple copies.

Under the current system, there's no complete listing of all movies in stock. The owner would like to be able to prepare a catalog of the entire collection based on information stored by the computer system. The catalog will be in alphabetical order by movie title and for each will contain a brief description of the film, whether it's available in VHS or Beta format (or both), and the number of copies in the collection.

In addition to a catalog of films, the owner would also like to be able to prepare listings of the entire inventory grouped by:

rating of the movie
type of movie (comedy, music, mystery, etc.)
year of release
title
tape format
individual cassette ID number

Also, the owner wants to be able to notify customers monthly of the new titles that are available. Each month, the system should be able to print a listing of just *new* titles. When customers visit the shop, they can take a copy of this listing.

Rental and return transactions Of course, if the system is to maintain information on customer and cassette activity, it must capture rental and return transaction data. So, terminals will be installed at the rental desk, and when a customer rents a movie, the clerk should have to enter only the customer's ID number and the ID numbers of the movies selected for rental in order to record the transaction.

A copy of the rental transaction must be printed on plain paper for the customer since no rental forms will be kept in the shop. The customer will no longer have to surrender his membership card because the computer system will keep track of how many cassettes he has on rental. If he has four on rental, the system won't allow him to rent any more.

When the member returns the movies he's rented, the clerk should be able to call up information on the original rental to calculate the amount due. All the clerk should have to do is enter the ID numbers of the returned cassettes. Then, the system should calculate the rental amount due and print a rental receipt on plain paper for the customer.

The system must issue reminders to customers who have overdue movies. This will relieve the clerks of having to make telephone calls to those customers. In addition, the system will print listings of customers with overdue movies in sequence by age (how long the tapes have been on rental) and member ID.

The system must also allow the shop's owner to revoke a customer's membership if he fails to return overdue movies. However, since revoking a customer's membership won't be an automatic function, the owner must make a terminal entry using information supplied by the system on the overdue reports.

Exercises

2-2-1 Identify the changes in functional requirements for the new video cassette rental system, as done in figure 4-1 in the text (page 80).

2-2-2 Identify the data flows that should be included on the model DFD for the new system. Simplify the system by omitting extraction processes (reporting and inquiry functions).

2-2-3 Create a model DFD for the new video cassette rental system using only the input and output data flows you identified in exercise 2-2-2.

Section 3: Data dictionary entries

2-3-1 Using data dictionary notation, create preliminary contents lists for the data stores on the model DFD for the new video cassette rental system you created in exercise 2-2-3. Include all of the data elements necessary to meet the reporting and inquiry requirements of the new system.

Section 4: System structure chart

2-4-1 Working from the model DFD, create a preliminary system structure chart for the new video cassette rental system.

2-4-2 Expand the processes on the preliminary system structure chart. To do so, use a table similar to the one in figure 6-3 of the text (page 127).

2-4-3 Add other required functions to the system structure chart that aren't part of the model DFD. If necessary, expand them to show their component functions. Be sure to group functions so the span of control of each module is reasonable. Number and name each module.

2-4-4 The new system will include terminals both at the rental desk and in the shop's office. All system functions will be available at the terminal in the office, but only those directly related to customer transactions should be available at the rental desk. Draw a system structure chart that depicts a menu structure that provides only the functions required at the rental desk.

Section 5: Database design

2-5-1 Identify the access paths to the data stores of the new video cassette rental system. Draw a data access diagram to show the relationships among the data stores and access paths of this system.

2-5-2 Why isn't it entirely appropriate to draw a data hierarchy diagram to show the relationships between the data stores of the new video cassette rental system?

2-5-3 Which of the access paths required by the system's data stores should be implemented by some keying mechanism in the system's file structure, and which should be simply sort keys? Why?

Section 6: Program specifications

2-6-1 Which modules on the system structure chart for the

new video cassette rental system are likely candidates for combination?

2-6-2 Prepare complete program specifications for the system structure chart module the clerk will use to enter rental transactions.

Be sure to include these components:

program overview
screen layouts
report layout for the rental form

Section 7: Program design

2-7-1 Using the program specifications from exercise 2-6-2, create a program structure chart for the enter-rental-transactions program.

Case Study 3

Vending machine management system

As a sideline business, a candy and tobacco wholesaler owns and operates several hundred vending machines stocked with items from his inventory. The vending machine business started as a secondary operation several years ago. Today, however, the operation has grown, and the owner has a substantial amount of money invested in the machines and employs five route men who service and stock them.

Although the firm uses a minicomputer system for its inventory, receivables, payables, and general ledger applications, record keeping for the vending machines is done manually. The owner believes he can keep better control of the vending machine operation if he takes advantage of the resources of his firm's computer system. In this case study, you'll develop a plan for doing that.

Section 1: Analysis DFD

The firm has 800 vending machines located throughout the community in 160 separate locations. Each unit is serviced once per week. To service a machine, a route driver collects the money in it and replaces sold items. Each of the five drivers services 30 to 35 machines each day.

Each morning, a route driver decides what products he needs that day to stock the machines he'll be servicing. He selects the stock he wants from the warehouse inventory and advises the inventory clerk of what he has taken.

The inventory clerk makes entries at a computer terminal to record the inventory depletion for the items the driver selected. (Remember, the inventory application is already a part of the firm's computer system.) From the terminal displays, the clerk determines the cost of the stock the driver selected, then records that cost in a journal for vending machine expenses.

After the driver has stocked the items needed for the day's rounds, he leaves on his route. At each stop, he restocks the machines and collects the money from them. At some locations, part of the money from the machine is turned over to the operator of the facility as a fee for allowing the machine on the premises. From other locations, the driver returns all the money collected to the office.

At the office, the bookkeeper accumulates the money received from each route driver and prepares a bank deposit at the end of each day. After the bank deposit is prepared, the bookkeeper enters the total collected in a journal for vending machine sales.

Once each month, the bookkeeper prepares a profitability report for the vending machine operation. To do so, he subtracts the period total in the expenses journal from the period total in the sales journal. The difference is gross profit for the vending machine operation.

When a new vending machine is purchased and installed, an identifying card is completed for it based on information supplied by the firm's accounts payable department. The card specifies the machine's manufacturer, type, serial number, and the number of different items it can contain.

The card also indicates where the machine is located, whom to contact at that location and at what telephone number, and the profit-splitting agreement for that location. If the machine is moved, that's also recorded on the card. The owner of the firm decides on occasion to move machines. When he does, he hires a moving company to do the job. After a machine is installed at a new location, the moving company notifies the firm.

The maintenance history of each machine is also kept on the identifying card. When maintenance work is done on a machine, the details are recorded on the back of the card. At the same time, the cost of the maintenance is recorded in the vending machine expenses journal. The information for maintenance data is taken from copies of paid maintenance bills supplied by the accounts payable department.

Exercises

3-1-1 Identify the inputs to and the outputs from the existing vending machine management system.

3-1-2 Draw the context DFD for the existing vending machine management system.

3-1-3 Identify the data stores of the vending machine management system.

3-1-4 Draw an analysis DFD for the existing vending machine management system.

Section 2: Model DFD

Because the firm has so much money invested in the machines, and because they require several employees to maintain them, the owner of the firm has decided it's important to keep more efficient and reliable records on the machines. To accomplish this, he wants to use his minicomputer system.

By improving record keeping for the machines, the firm's owner hopes that he'll be able to (1) keep better control of stocking the machines and collecting money from them and (2) make better decisions about where to locate the machines.

The owner wants better control over stocking the machines because he believes some of his route men are stealing inventory by taking more than they need to stock the machines. Now, however, he has no way of telling whether or not that's true because he can't identify the stock each driver needs to service the machines on his route.

Worse than possible theft of inventory items, the owner of the firm is unable to tell if his route drivers are turning in the total amount of money they collect from the machines they service. During several accounting periods, the vending machine operation has actually lost money, and the owner is convinced it's because of theft.

Beyond controlling the inventory and cash handling associated with the vending machine operation, the owner would like to have information that will help him decide if it's worth keeping machines at particular locations. The information he needs to do this is sales data by location: cumulative, year-to-date, and month-to-date.

Exercises

3-2-1 Identify the changes in functional requirements for the new vending machine management system.

3-2-2 The solution to this problem isn't as obvious from the case description as in the previous case studies. Although the functional requirements of the new system are simple, you as the analyst/designer have great flexibility in developing a solution. Consider alternate ways to implement the functional requirements of the new system. Focus on the system inputs and outputs that will meet the owner's needs. What are the apparent cost advantages and disadvantages of alternate implementations? Be prepared to discuss them.

3-2-3 Create a model DFD for the new vending machine management system. Indicate the boundary of the new computer system on the model DFD.

Section 3: Data dictionary entries

3-3-1 Using data dictionary notation, create preliminary contents lists for the data stores on the model DFD for the new vending machine management system.

Section 4: System structure chart

3-4-1 Working from the model DFD, create a preliminary system structure chart for the new vending machine management system. If you've shown extraction processes on your model DFD, include them on the structure chart, but shade them to indicate you don't need to analyze them in the next exercise.

3-4-2 Expand the operational processes on the preliminary system structure chart (not the extraction processes you shaded). Use a table similar to the one in figure 6-3 of the text (page 127). Then, redraw the system structure chart to include the new functions you've identified. Group the modules logically, and name and number each of them.

3-4-3 This system must produce profitability, audit, and location sales reports at the beginning of each month. In addition, the total fields in the system's data stores must be rolled forward at the beginning of each month. How should the top-level module of the system structure chart work to insure that these events will occur at the right time?

Section 5: Database design

3-5-1 Identify the access paths to the data stores of the new vending machine system. Draw a data access diagram to show the relationships among the data stores and access paths of this system.

3-5-2 The relationships shown in the data access diagram you developed in exercise 3-5-1 should suggest a hierarchical structure for the system's data stores. Draw a data hierarchy diagram to show that structure. Then, rename the data stores to more accurately reflect their position in the hierarchy.

Section 6: Program specifications

3-6-1 Which modules on the system structure chart for the new vending machine management system are likely candidates for combination?

3-6-2 Prepare a print chart for the route list the driver will use to record cash collections.

3-6-3 Prepare complete program specifications for the system structure chart module that will be used to enter the cash collected by the route driver.

Be sure to include these components:

program overview
screen layouts
report layout for the audit trail document the program should create

Section 7: Program design

3-7-1 Using the program specifications from exercise 1-6-2, create a program structure chart for the enter-cash-receipts program.

Case Study 4

Brokerage system reporting enhancement

When I developed and installed the brokerage system the text describes, one of the output items I included was a position report. This report compares the amount of a specified commodity purchased with the amount sold. The firm's brokers use the report to balance their transactions so they don't end up long (with more of a commodity purchased than sold) or short (more sold than purchased).

Figure 4-1 is a sample of the position report. As you can see, the report is prepared for a specified commodity (in this case, barley) and consists of two vertical sections: the left-hand section is for purchases, the right-hand section is for sales. Within these two sections, one line is printed for each open contract. The data printed is customer code, contract number, and quantity of the commodity undelivered.

As the original system was being implemented, the users thought this report would be adequate, and it is for most commodities. For a few, however, more detailed information is needed.

The brokerage trades some commodities, particularly grains, well in advance of when they'll actually be delivered. The problem with the current position report is that it doesn't indicate when the commodities it lists are to be delivered.

For example, the report in figure 4-1 shows the brokerage long on the commodity barley by about 10,973 tons (21,946,281 pounds). So it appears the brokerage will be able to meet its sale obligations easily. The fact is, though, that's not the case. Most of the purchases shown on the report are actually due for delivery in June, 1984. The sale delivery requirements, however, are more evenly distributed on a monthly basis, January through June. The result is that the brokerage is short for February, March, April, and May. And it's very long for June. If the brokers depended entirely on the position report in figure 4-1, it wouldn't be long before they faced serious problems meeting their contract obligations.

To make the system more useful for the traders, the brokerage's management wants to modify the system so it can keep track of and report on future delivery obligations. In this case study, you'll plan the modifications to the system necessary to meet these requirements.

The current position report is prepared in a two-step process. First, the two contract files, named SALECON and PURCHCON, are sorted into sequence by contract within customer within commodity. (That's the sequence in which the contracts appear on the report in figure 4-1.)

```
DATE: 01/31/84                                                                                   PAGE: 2
TIME: 12:24 PM                                                                                   MGT2100
                                        P O S I T I O N   R E P O R T

*****************************************************************************************************
COMMODITY: B       (BARLEY)
*****************************************************************************************************

                  P U R C H A S E S                                          S A L E S

     SELLER            CONTRACT            UNDELIVERED          BUYER          CONTRACT          UNDELIVERED

     MILLER, S. J.     01455-0              320,000  LBS
     NEWELL            01663-0            1,800,000  LBS
     NEWT              01607-0              800,000  LBS
     OLIVERA, M.       01605-0               60,000  LBS
     ORLANDO BROS.     01686-0              600,000  LBS
     PACHECO           01457-0              600,000  LBS
     PEAVEY            01602-0              900,000  LBS
     PHILLIPS INDUST   01685-0              180,000  LBS
     PINHEIRO, S. C.   01635-0              400,000  LBS
     PINOLE            01573-0              110,000  LBS
     PLK               01598-0            1,100,000  LBS
     PRYSE FARMS       01599-0              660,000  LBS
     PUCHEU BROS.      01683-0            1,000,000  LBS
     REECELANDS WEST   01676-0            2,400,000  LBS
     ROMEIRO RANCH     01638-0              200,000  LBS
     SCOULAR GRAIN     01570-0              774,100  LBS
                       01572-0              277,240  LBS
     SILVEIRA, S.      01114-0              300,000  LBS
     SINGHRAI, N.      01642-0              500,000  LBS
     SMITH, Z.         01456-0              400,000  LBS
                       01600-0            1,000,000  LBS
     SODA SPRINGS      01501-0               36,600  LBS
     TOOMEY, M.        01621-0              260,000  LBS
     VELHUIZEN, A.     01409-0              400,000  LBS
     VERBOON           01611-0              200,000  LBS
     VIERHUS           00961-0            2,000,000  LBS
     VIERRA, E.        01633-0              100,000  LBS
     WHITFIELD, L.     01687-0              500,000  LBS
     WOODS, P.         01677-0            4,400,000  LBS
     YAGER             01680-0              360,000  LBS
     ZANDUETTA         01678-0              500,000  LBS
                                     ------------------                                  -----------------
                                         47,725,650  LBS                                     25,779,369  LBS
*****************************************************************************************************
```

Figure 4-1 (Part 2 of 2) The existing position report

DATE: 01/31/84
TIME: 12:24 PM

PAGE: 1
MGT2100

POSITION REPORT

COMMODITY: B (BARLEY)

PURCHASES				SALES			
SELLER	CONTRACT	UNDELIVERED		BUYER	CONTRACT	UNDELIVERED	
AGRI BEEF	01612-0	360,000	LBS	BARK	01663-0	2,000,000	LBS
ALTA GROWERS	01462-0	500,000	LBS	COAST	00731-0	1,879,580	LBS
AYERZA G.	01592-0	300,000	LBS		00745-0	1,404,640	LBS
BAKER T.	01492-0	110,000	LBS		00776-0	2,791,040	LBS
	01668-0	100,000	LBS		00864-0	2,180,820	LBS
BARLOW	01601-0	1,300,000	LBS		01342-0	1,851,520	LBS
BBZ	01636-0	280,000	LBS	DCCA	01802-0	1,800,000	LBS
	01637-0	280,000	LBS	JDH	00608-0	2,159,410	LBS
	01640-0	280,000	LBS		00613-0	2,394,210	LBS
BEN & ESP	01670-0	150,000	LBS		00787-0	4,841,469	LBS
BENEVEDES	01669-0	110,000	LBS	RALSTEN	01811-0	180,000	LBS
BERRY R.	01225-0	950,000	LBS	VERH F	01709-0	1,000,000	LBS
CARGILL-S	00486-0	93,560	LBS	WES CON-C	01771-0	296,680	LBS
CARL, J.	01610-0	400,000	LBS		01773-0	1,000,000	LBS
CHANGALA, A.	01608-0	600,000	LBS				
CHANGALA, S.	01609-0	300,000	LBS				
COMBS, M.	01671-0	500,000	LBS				
COOPER	01630-0	400,000	LBS				
DAVIS, M.	01625-0	640,000	LBS				
DIAMOND R	01672-0	400,000	LBS				
E C	01639-0	1,300,000	LBS				
FARM CO-OP	00441-0	1,980,000	LBS				
	00444-0	2,160,000	LBS				
FLY TIG	01503-0	160,000	LBS				
FOWLER	01143-0	1,400,000	LBS				
FRANCINI, D.	01626-0	160,000	LBS				
GATEWAY	01664-0	900,000	LBS				
GUNN, L.	01673-0	150,000	LBS				
GVF-#2	01684-0	1,700,000	LBS				
HANSEN, W.	01148-0	1,000,000	LBS				
HARRELL	01463-0	100,000	LBS				
HOFFER	01590-0	200,000	LBS				
HUGHES BROS.	01589-0	300,000	LBS				
IDAHO GRAIN	01595-0	560,000	LBS				
LEES	00451-0	200,150	LBS				
MARCH BROS.	01675-0	1,800,000	LBS				
MARTELLA, B.	01632-0	100,000	LBS				
MARTIN, JOE	01622-0	224,000	LBS				
MATHIAS, R.	01681-0	280,000	LBS				
MELCOMBS	01674-0	860,000	LBS				
MENDIBORU, J.	01546-0	1,000,000	LBS				

Figure 4-1 (Part 1 of 2) The existing position report

Then, once the files have been sorted, the report-preparation program is run. It's an interactive program that displays the screen in figure 4-2. The broker requesting the report keys in the code for the commodity for which he wants information. The program validates the code by using it as the primary key value for a read on the reference file COMMODTY. If the selected code is valid, the program reads through both sorted contract files sequentially until it reaches the first record in each for the specified commodity. The program prints a line for each open purchase and sale contract for the requested commodity. The quantity undelivered is calculated by subtracting the delivered quantity for each contract from the original contract quantity. The program accumulates the total undelivered purchase and sale quantities for the commodity and, when the last contracts for the commodity have been processed, prints those totals.

The new future position report must show for each contract the quantity of commodity to be delivered in the current month and in the 18 months following the current month. At the end of the report, the quantity purchased, the quantity sold, and the difference (long or short) must be printed for the current month and for each of the following 18 months. When all contract records for the specified commodity have been read, a grand total line for all periods should be printed (this line should contain the same data as the total line on the current position report).

The future-position-report program should work in the same way as the existing program. Unfortunately, the existing contract record formats don't provide for future period delivery quantities. As a result, some changes will have to be made to the system's files and programs to allow that information to be stored.

Keep in mind that the future delivery reporting requirement applies to a relatively small proportion of all contracts the system maintains. Also, remember that the purchase and sale contract files and the commodity file are used by almost all of the programs that make up the system. To modify their format could mean substantial programming changes.

Figures 4-3 through 4-9 give you the system documentation that you may need for this case study. This includes everything from the model DFD for the brokerage system to the COBOL COPY members for the record descriptions for the sale-contracts, purchase-contracts, and commodity-reference files.

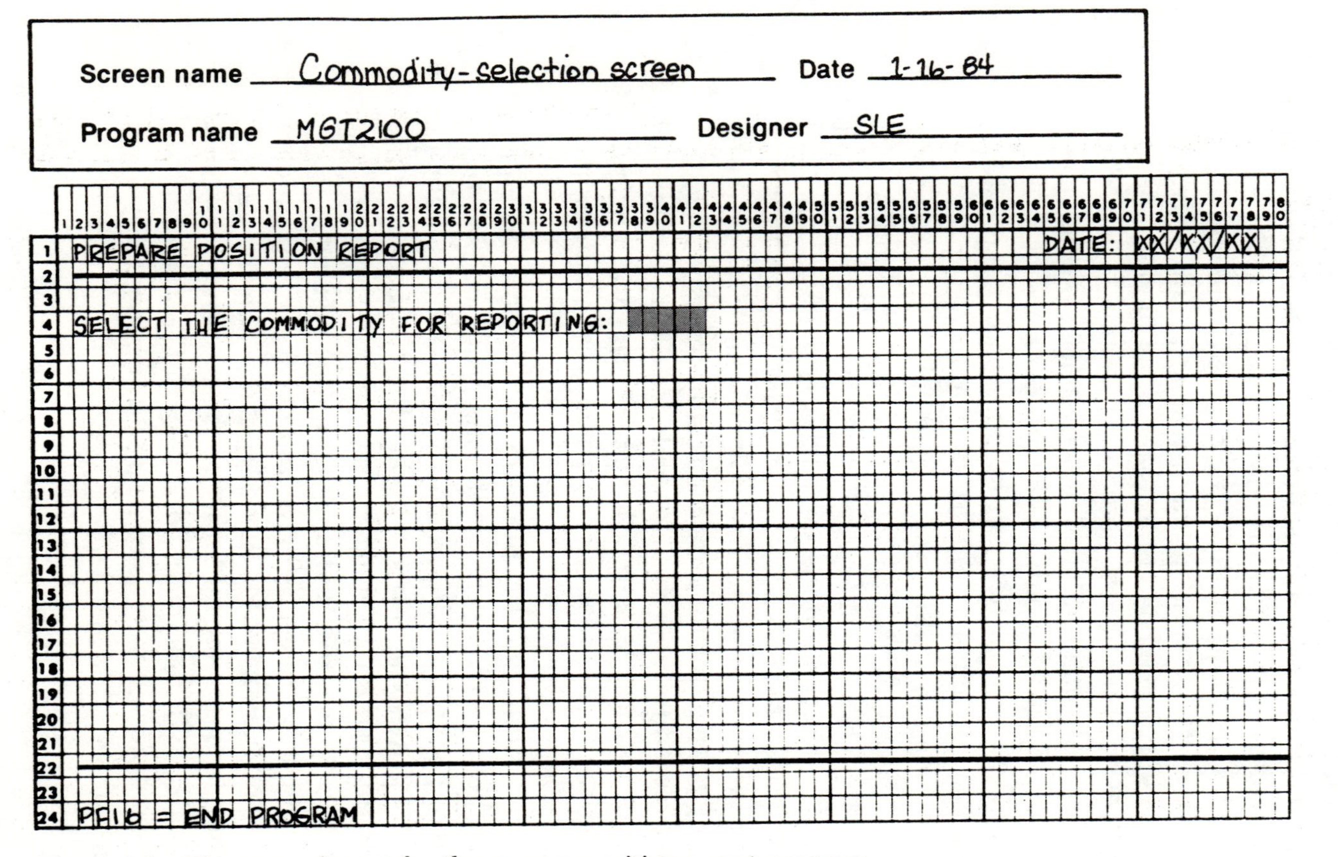

Figure 4-2 The screen layout for the prepare-position-report program

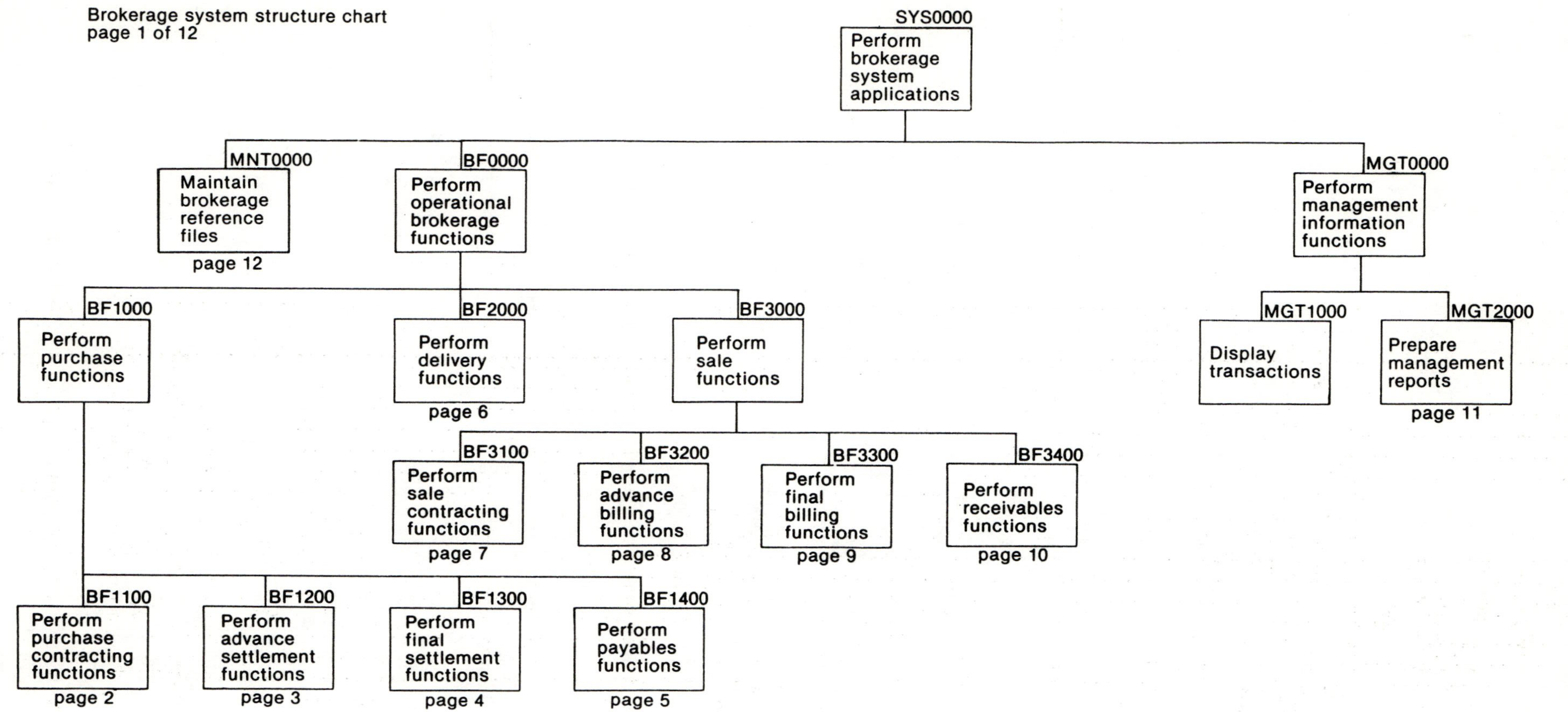

Figure 4-4 (Part 1 of 12) The refined system structure chart for the brokerage system (figure 6-8 in the text)

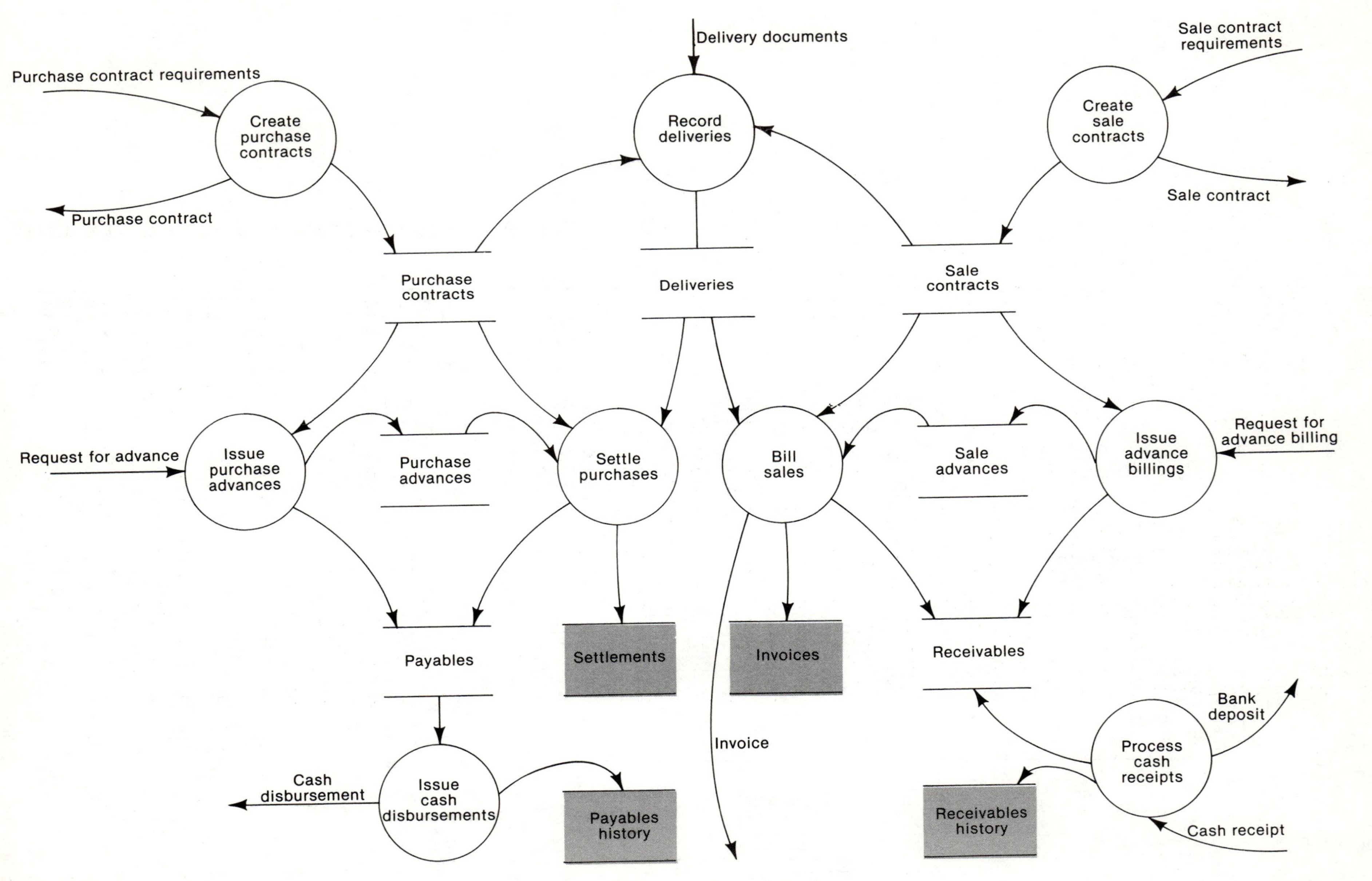

Figure 4-3 The model DFD for the brokerage system showing the addition of the archive data stores (figure 7-4 in the text)

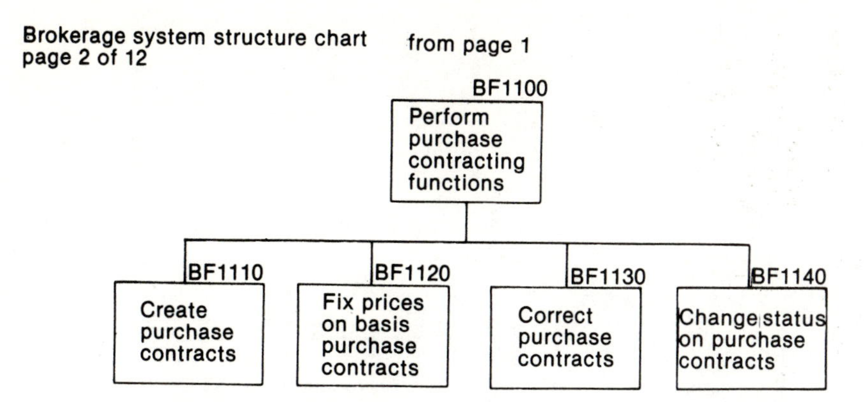

Figure 4-4 (Part 2 of 12) The refined system structure chart for the brokerage system (figure 6-8 in the text)

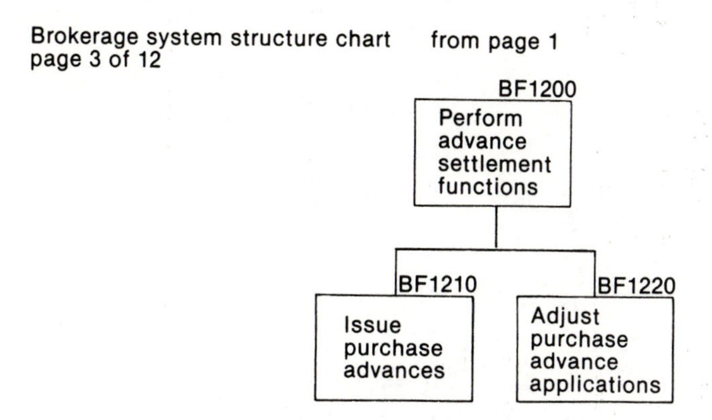

Figure 4-4 (Part 3 of 12) The refined system structure chart for the brokerage system (figure 6-8 in the text)

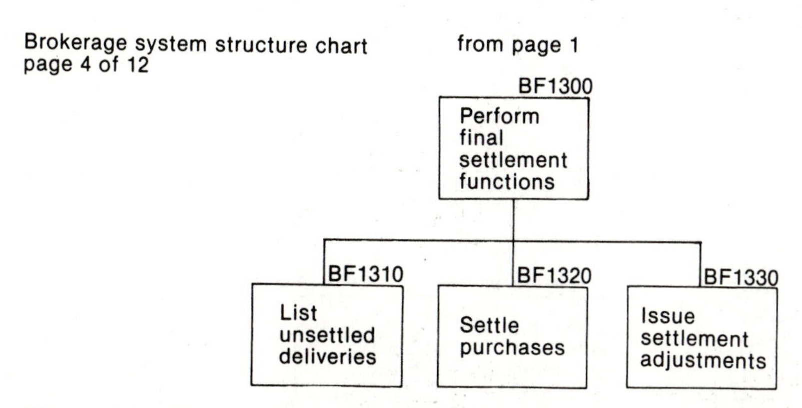

Figure 4-4 (Part 4 of 12) The refined system structure chart for the brokerage system (figure 6-8 in the text)

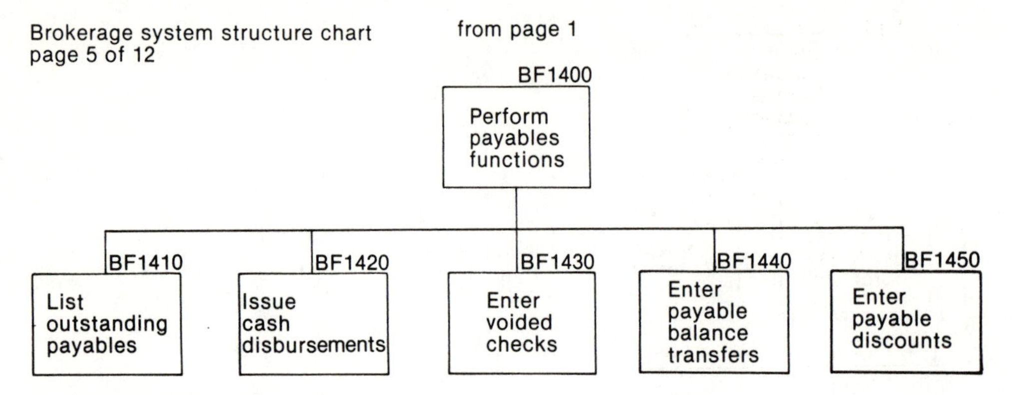

Figure 4-4 (Part 5 of 12) The refined system structure chart for the brokerage system (figure 6-8 in the text)

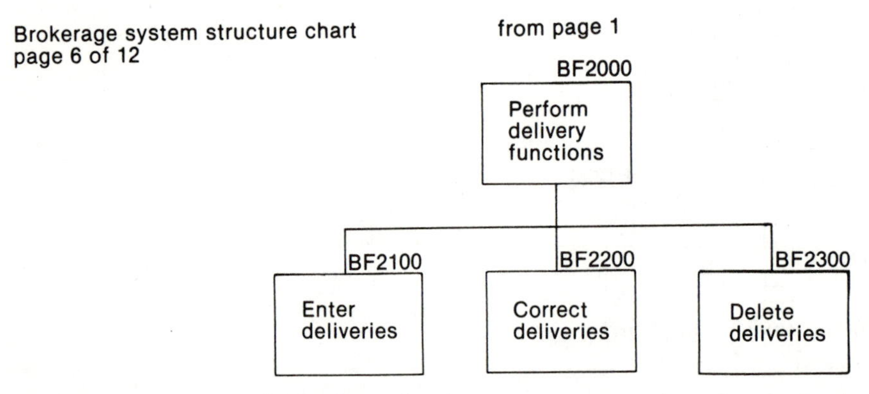

Figure 4-4 (Part 6 of 12) The refined system structure chart for the brokerage system (figure 6-8 in the text)

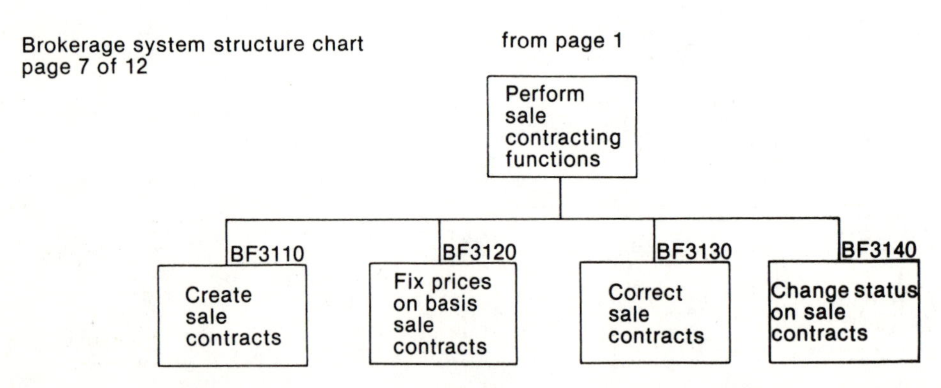

Figure 4-4 (Part 7 of 12) The refined system structure chart for the brokerage system (figure 6-8 in the text)

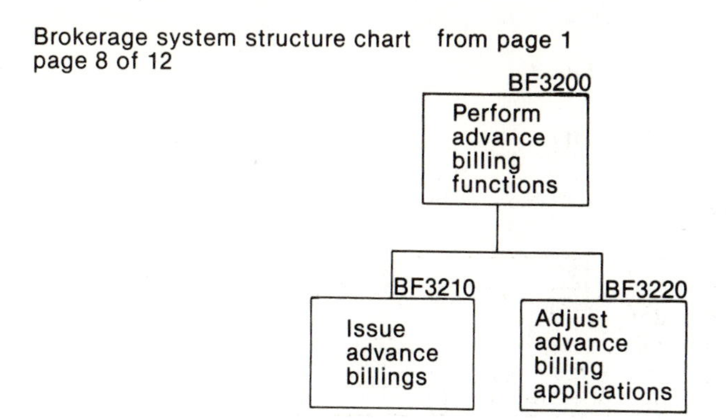

Figure 4-4 (Part 8 of 12) The refined system structure chart for the brokerage system (figure 6-8 in the text)

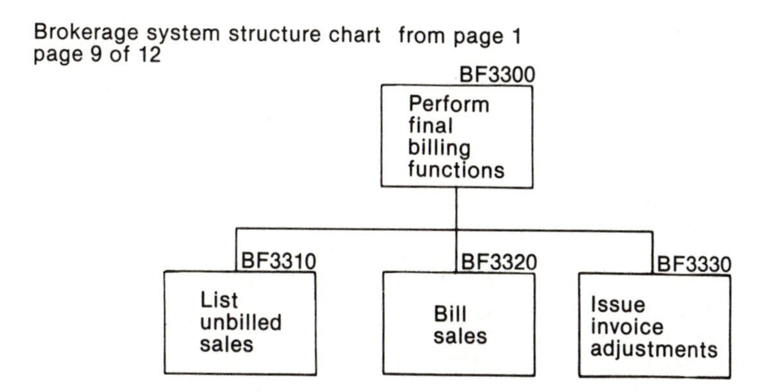

Figure 4-4 (Part 9 of 12) The refined system structure chart for the brokerage system (figure 6-8 in the text)

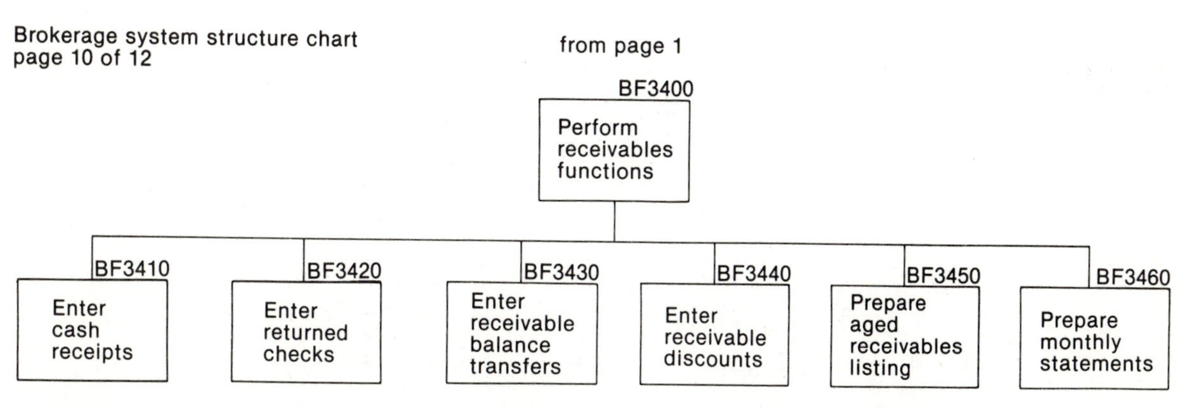

Figure 4-4 (Part 10 of 12) The refined system structure chart for the brokerage system (figure 6-8 in the text)

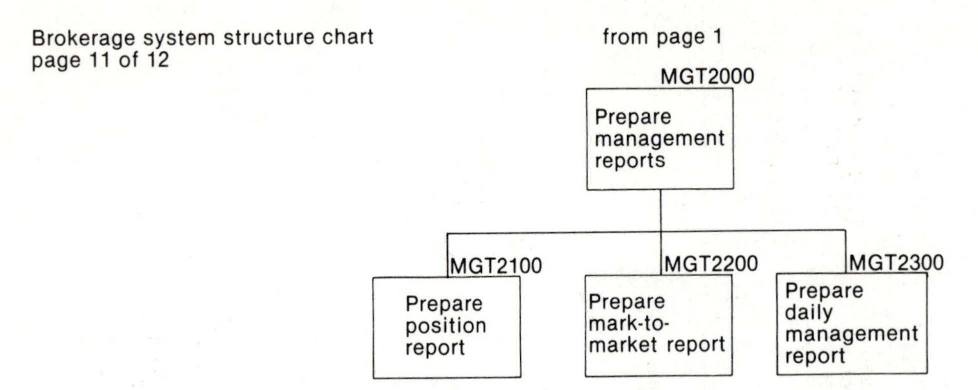

Figure 4-4 (Part 11 of 12) The refined system structure chart for the brokerage system (figure 6-8 in the text)

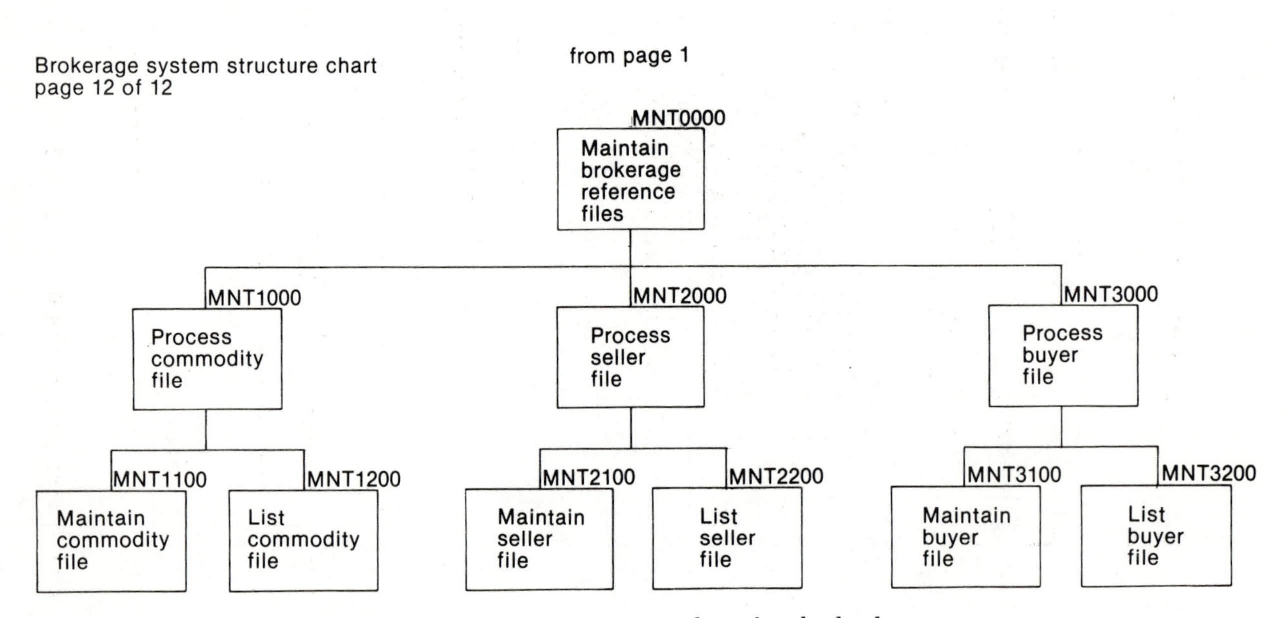

Figure 4-4 (Part 12 of 12) The refined system structure chart for the brokerage system (figure 6-8 in the text)

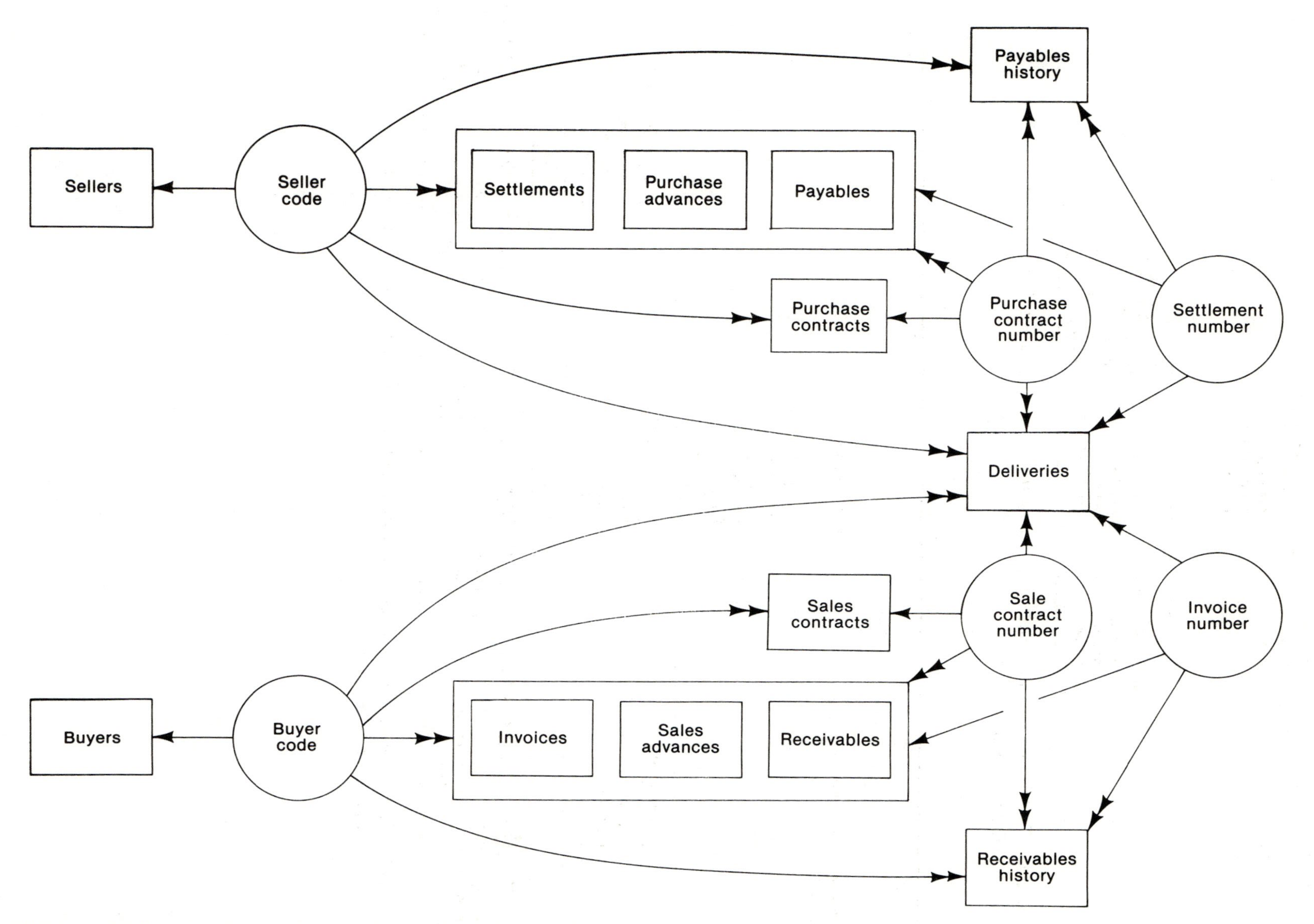

Figure 4-5 The data access diagram for the brokerage system (figure 7-8 in text)

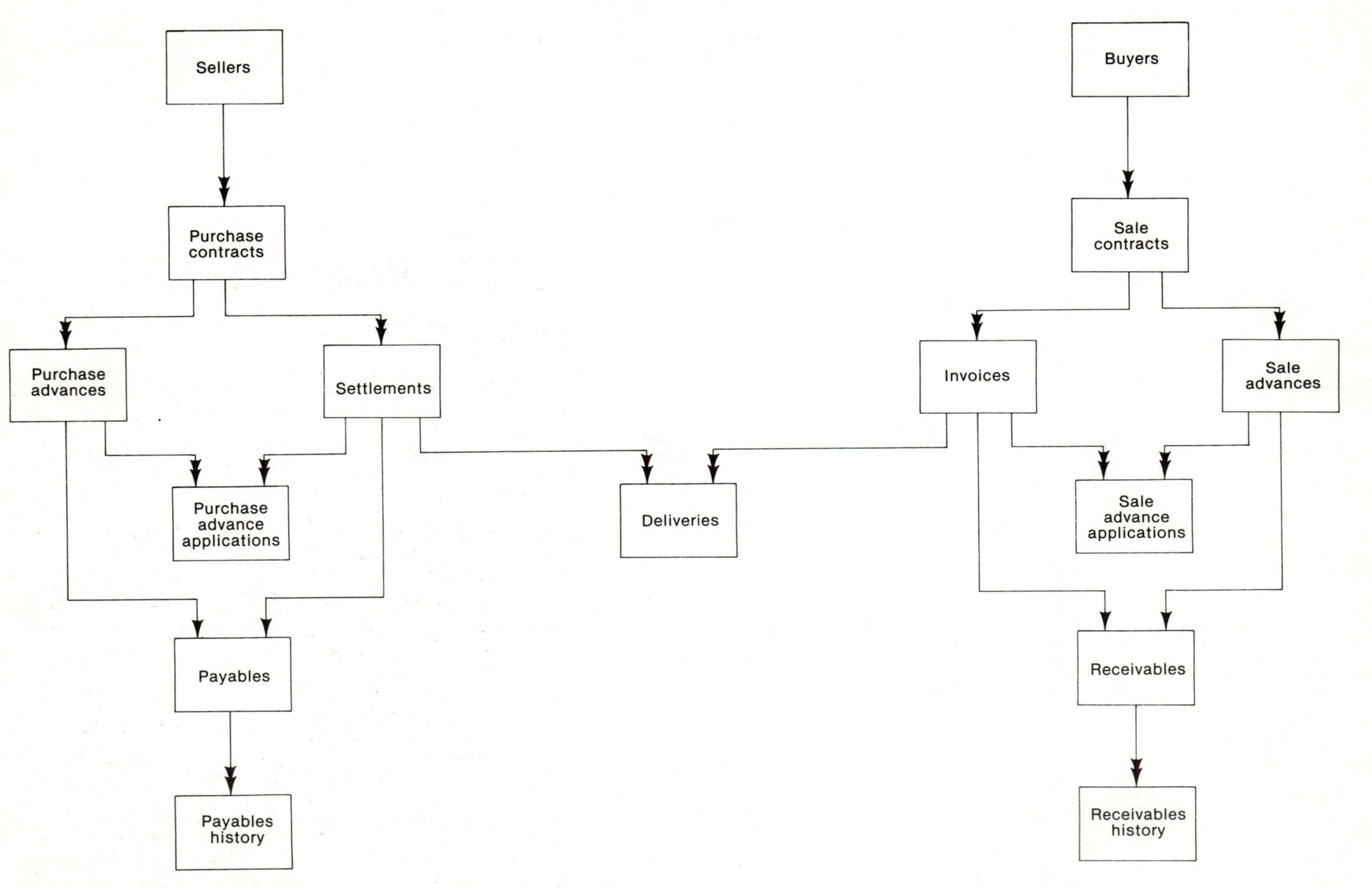

Figure 4-6 The data hierarchy diagram for the brokerage system (figure 7-11 in text)

```
 01   SALE-CONTRACT-RECORD.
*
     05   SCR-CONTRACT-KEY.
          10   SCR-CONTRACT-NUMBER          PIC 9(5).
          10   SCR-CONTRACT-SEQ-NUMBER      PIC 9.
     05   SCR-CONTRACT-DATE.
          10   SCR-CONTRACT-MONTH           PIC XX.
          10   SCR-CONTRACT-DAY             PIC XX.
          10   SCR-CONTRACT-YEAR            PIC XX.
     05   SCR-BUYER-CODE                    PIC X(16).
     05   SCR-COMMODITY-DATA.
          10   SCR-COMMODITY-CODE           PIC X(5).
          10   SCR-COMMODITY-VARIETY        PIC X(30).
          10   SCR-COMMODITY-QUALITY        PIC X(30).
          10   SCR-COMMODITY-COMMENT        PIC X(30).
     05   SCR-QUANTITY                      PIC S9(9)    COMP.
     05   SCR-MEASURING-UNIT                PIC X.
          88   SCR-BUSHELS                               VALUE "B".
          88   SCR-CWTS                                  VALUE "H".
          88   SCR-POUNDS                                VALUE "P".
          88   SCR-TONS                                  VALUE "T".
          88   SCR-EACH                                  VALUE "E".
     05   SCR-SHIPMENT-DATA.
          10   SCR-SHIPMENT-METHOD-FLAG     PIC X.
               88   SCR-FOB-TRUCK                        VALUE "T".
               88   SCR-FOB-RAIL                         VALUE "R".
               88   SCR-FOB-EITHER-TRUCK-OR-RAIL         VALUE "E".
          10   SCR-SHIPMENT-DESTINATION     PIC X(30).
          10   SCR-SHIPMENT-DATE.
               15   SCR-SHIPMENT-MONTH      PIC XX.
               15   SCR-SHIPMENT-DAY        PIC XX.
               15   SCR-SHIPMENT-YEAR       PIC XX.
     05   SCR-DELIVERIES-COMPLETED-SW       PIC X.
          88   SCR-DELIVERIES-COMPLETED                  VALUE "Y".
     05   SCR-PRICE-DATA.
          10   SCR-PRICE-TYPE               PIC X.
               88   SCR-BASIS-PRICING                    VALUE "B".
               88   SCR-FIXED-PRICING                    VALUE "F".
          10   SCR-PRICE                    PIC S9(3)V9999  COMP.
          10   SCR-PRICING-UNIT             PIC X.
               88   SCR-PRICE-UNIT-BUSHELS               VALUE "B".
               88   SCR-PRICE-UNIT-CWTS                  VALUE "H".
               88   SCR-PRICE-UNIT-POUNDS                VALUE "P".
               88   SCR-PRICE-UNIT-TONS                  VALUE "T".
               88   SCR-PRICE-UNIT-EACH                  VALUE "E".
          10   SCR-MARKET-PRICE-WHEN-FIXED PIC S9(3)V9999  COMP.
               10   SCR-MKT-PRICING-UNIT    PIC X.
               88   SCR-MKT-PRICE-UNIT-BUSHELS           VALUE "B".
               88   SCR-MKT-PRICE-UNIT-CWTS              VALUE "H".
               88   SCR-MKT-PRICE-UNIT-POUNDS            VALUE "P".
               88   SCR-MKT-PRICE-UNIT-TONS              VALUE "T".
               88   SCR-MKT-PRICE-UNIT-EACH              VALUE "E".
          10   SCR-BASIS-PRICE-DATA.
               15   SCR-BASIS-PRICING-UNIT  PIC X.
                    88   SCR-BASIS-PRICE-UNIT-BUSHELS    VALUE "B".
```

Figure 4-7   (Part 1 of 2) The record description for the file SALECON presented as a COBOL copy member

```
                88   SCR-BASIS-PRICE-UNIT-CWTS          VALUE "H".
                88   SCR-BASIS-PRICE-UNIT-POUNDS        VALUE "P".
                88   SCR-BASIS-PRICE-UNIT-TONS          VALUE "T".
                88   SCR-BASIS-PRICE-UNIT-EACH          VALUE "E".
            15  SCR-BASIS-OVER-OR-UNDER PIC X.
                88   SCR-BASIS-OVER                     VALUE "O".
                88   SCR-BASIS-UNDER                    VALUE "U".
            15  SCR-BASIS-DATE.
                20   SCR-BASIS-MONTH       PIC XX.
                20   SCR-BASIS-YEAR        PIC XX.
            15  SCR-BASIS-CITY             PIC X.
                88   SCR-CHICAGO                        VALUE "C".
                88   SCR-KANSAS-CITY                    VALUE "K".
            15  SCR-BASIS-COMMODITY        PIC X(5).
    05  SCR-POUNDS-DELIVERED               PIC S9(13)   COMP.
    05  SCR-DOCKAGE-ALLOWABLE-FLAG         PIC X.
        88   SCR-OVER-1-DEDUCTIBLE                      VALUE "1".
        88   SCR-ALL-DEDUCTIBLE                         VALUE "A".
        88   SCR-NO-DOCKAGE                             VALUE "N".
*
```

Figure 4-7 (Part 2 of 2) The record description for the file SALECON presented as a COBOL copy member

```
 01  PURCHASE-CONTRACT-RECORD.
*
     05  PCR-CONTRACT-KEY.
         10  PCR-CONTRACT-NUMBER             PIC 9(5).
         10  PCR-CONTRACT-SEQ-NUMBER         PIC 9.
     05  PCR-CONTRACT-DATE.
         10  PCR-CONTRACT-MONTH              PIC XX.
         10  PCR-CONTRACT-DAY                PIC XX.
         10  PCR-CONTRACT-YEAR               PIC XX.
     05  PCR-SELLER-CODE                     PIC X(16).
     05  PCR-COMMODITY-DATA.
         10  PCR-COMMODITY-CODE              PIC X(5).
         10  PCR-COMMODITY-VARIETY           PIC X(30).
         10  PCR-COMMODITY-QUALITY           PIC X(30).
         10  PCR-COMMODITY-COMMENT           PIC X(30).
     05  PCR-QUANTITY                        PIC S9(9)    COMP.
     05  PCR-MEASURING-UNIT                  PIC X.
         88  PCR-BUSHELS                                  VALUE "B".
         88  PCR-CWTS                                     VALUE "H".
         88  PCR-POUNDS                                   VALUE "P".
         88  PCR-TONS                                     VALUE "T".
         88  PCR-EACH                                     VALUE "E".
     05  PCR-SHIPMENT-DATA.
         10  PCR-SHIPMENT-METHOD-FLAG        PIC X.
             88  PCR-FOB-TRUCK                            VALUE "T".
             88  PCR-FOB-RAIL                             VALUE "R".
             88  PCR-FOB-EITHER-TRUCK-OR-RAIL             VALUE "E".
         10  PCR-SHIPMENT-ORIGIN             PIC X(30).
         10  PCR-SHIPMENT-DATE               PIC X(30).
             15  PCR-SHIPMENT-MONTH          PIC XX.
             15  PCR-SHIPMENT-DAY            PIC XX.
             15  PCR-SHIPMENT-YEAR           PIC XX.
     05  PCR-DELIVERIES-COMPLETED-SW         PIC X.
         88  PCR-DELIVERIES-COMPLETED                     VALUE "Y".
     05  PCR-PRICE-DATA.
         10  PCR-PRICE-TYPE                  PIC X.
             88  PCR-BASIS-PRICING                        VALUE "B".
             88  PCR-FIXED-PRICING                        VALUE "F".
         10  PCR-PRICE                       PIC S9(3)V9999  COMP.
         10  PCR-PRICING-UNIT                PIC X.
             88  PCR-PRICE-UNIT-BUSHELS                   VALUE "B".
             88  PCR-PRICE-UNIT-CWTS                      VALUE "H".
             88  PCR-PRICE-UNIT-POUNDS                    VALUE "P".
             88  PCR-PRICE-UNIT-TONS                      VALUE "T".
             88  PCR-PRICE-UNIT-EACH                      VALUE "E".
         10  PCR-MARKET-PRICE-WHEN-FIXED PIC S9(3)V9999  COMP.
             10  PCR-MKT-PRICING-UNIT     PIC X.
             88  PCR-MKT-PRICE-UNIT-BUSHELS               VALUE "B".
             88  PCR-MKT-PRICE-UNIT-CWTS                  VALUE "H".
             88  PCR-MKT-PRICE-UNIT-POUNDS                VALUE "P".
             88  PCR-MKT-PRICE-UNIT-TONS                  VALUE "T".
             88  PCR-MKT-PRICE-UNIT-EACH                  VALUE "E".
         10  PCR-BASIS-PRICE-DATA.
             15  PCR-BASIS-PRICING-UNIT  PIC X.
                 88  PCR-BASIS-PRICE-UNIT-BUSHELS         VALUE "B".
```

Figure 4-8 (Part 1 of 2) The record description for the file PURCHCON presented as a COBOL copy member

```
                88   PCR-BASIS-PRICE-UNIT-CWTS              VALUE "H".
                88   PCR-BASIS-PRICE-UNIT-POUNDS            VALUE "P".
                88   PCR-BASIS-PRICE-UNIT-TONS              VALUE "T".
                88   PCR-BASIS-PRICE-UNIT-EACH              VALUE "E".
            15  PCR-BASIS-OVER-OR-UNDER PIC X.
                88   PCR-BASIS-OVER                         VALUE "O".
                88   PCR-BASIS-UNDER                        VALUE "U".
            15  PCR-BASIS-DATE.
                20   PCR-BASIS-MONTH       PIC XX.
                20   PCR-BASIS-YEAR        PIC XX.
            15  PCR-BASIS-CITY             PIC X.
                88   PCR-CHICAGO                            VALUE "C".
                88   PCR-KANSAS-CITY                        VALUE "K".
            15  PCR-BASIS-COMMODITY        PIC X(5).
        05  PCR-POUNDS-DELIVERED           PIC S9(13)       COMP.
        05  PCR-DOCKAGE-ALLOWABLE-FLAG     PIC X.
            88   PCR-OVER-1-DEDUCTIBLE                      VALUE "1".
            88   PCR-ALL-DEDUCTIBLE                         VALUE "A".
            88   PCR-NO-DOCKAGE                             VALUE "N".
```

Figure 4-8 (Part 2 of 2) The record description for the file PURCHCON presented as a COBOL copy member

```
 01   COMMODITY-RECORD.
*
      05   CM-COMMODITY-CODE                   PIC X(5).
      05   CM-COMMODITY                        PIC X(25).
      05   CM-VARIETY                          PIC X(25).
      05   CM-QUALITY                          PIC X(25).
      05   CM-POUNDS-PER-BUSHEL                PIC S9(3) COMP.
```

Figure 4-9 The record description for the file COMMODTY presented as a COBOL copy member

Section 1: Analysis DFD

4-1-1 Identify the inputs to and the outputs from the segment of the brokerage system that prepares the current position report.

4-1-2 Draw an analysis DFD for the segment of the brokerage system that prepares the current position report.

Section 2: Model DFD

4-2-1 Identify the inputs and outputs required to prepare the future position report.

4-2-2 Draw a model DFD for the segment of the brokerage system that will prepare the future position report.

Section 3: Data dictionary entries

4-3-1 Use data dictionary notation to create preliminary contents lists for the data stores on the model DFD that aren't part of the existing brokerage system.

Section 4: System structure chart

4-4-1 Identify the new programs that will be required to implement the preparation of the future position report. Assign names and numbers to them so they fit properly into the system structure chart in figure 4-4.

4-4-2 How can you insure that the future period data the system maintains corresponds to the correct time period? In other words, when the month changes, how will the system roll the temporal delivery data forward?

Section 5: Database design

4-5-1 Draw a data access diagram and a data hierarchy diagram for the data stores on the model DFD for the future position report.

4-5-2 Should the future delivery data be part of the contract records, both for purchases and sales? Why or why not?

Section 6: Program specifications

4-6-1 Create a print chart for the future position report.

4-6-2 Create complete program specifications for the program that will allow an operator to enter future delivery schedules for purchase contracts.

Be sure to include a program overview and screen layouts. No print output is required.

Section 7: Program design

4-7-1 Create a program structure chart from the specifications developed in exercise 4-6-2.

Appendix A

A summary of the procedures for analysis, design, and implementation

Analysis

1. Create an analysis DFD for the existing system.
    - Create the context analysis DFD.
    - Create the analysis DFD.
    - If necessary, simplify the analysis DFD.

2. Create a model DFD for the new system.
    - Identify the changes in functional requirements for the new system.
    - Establish the context for the new system.
    - Create a context DFD for the new system.
    - Create the model DFD for the new system.

Design

3. Define the data requirements.
    - Create preliminary contents lists for the system's data flows.

4. Create a system structure chart.
    - Create the preliminary chart.
    - Expand the original processes.
    - Add other required processes.
    - Expand the other processes.

5. Design the database.
    - Identify omitted data stores.
    - Draw a data access diagram to show required access paths.
    - Draw a data hierarchy diagram to clarify data relationships.
    - Evaluate and refine the database plan.
    - Decide how to package the system's data elements.
    - Document the physical database design.

6. Create the program specifications.
    - Identify useful standardized code.
    - Decide how to package the system modules into programs.
    - Design the programs' reports and screens and write a program overview.

Implementation

7. Develop the program of the system using structured programming techniques.
    - Create a program structure chart.
    - Implement the program using structured coding techniques and top-down testing.

8. Document the system.

Appendix B

Form masters

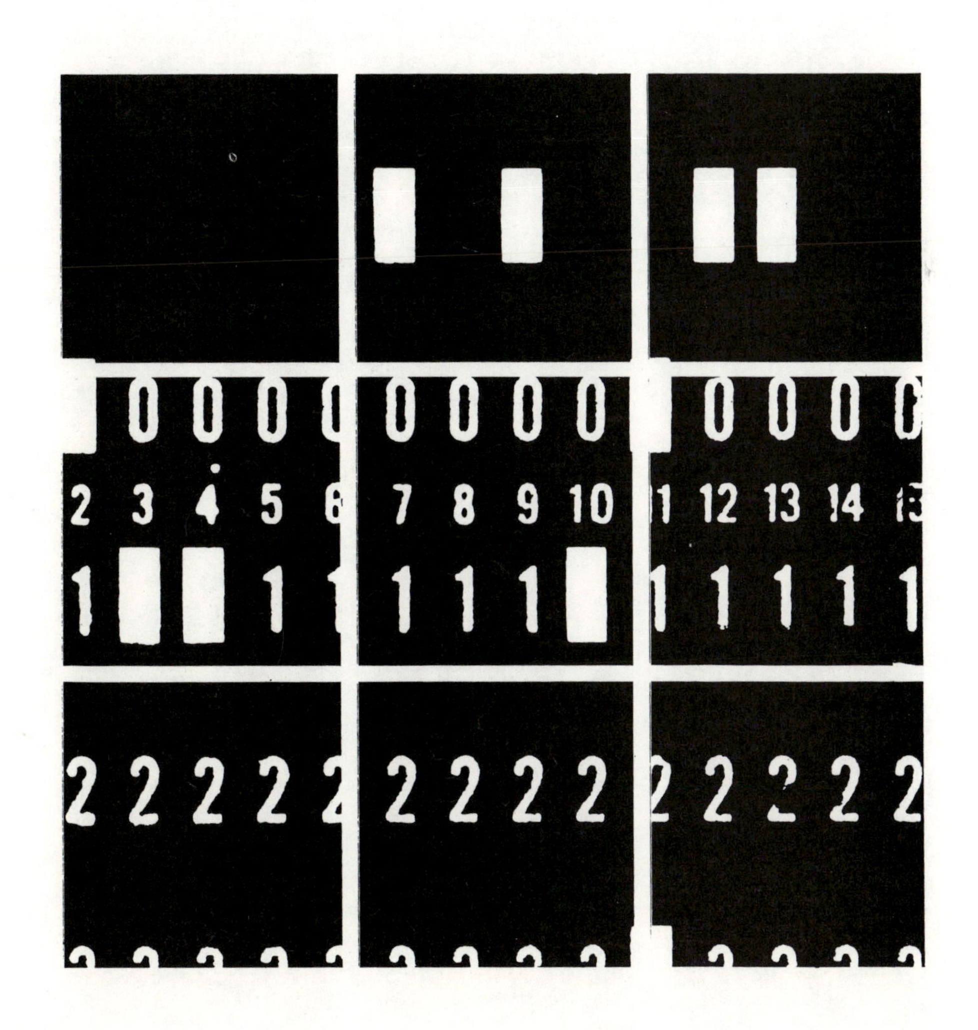

In this section, you'll find blank master copies of some of the forms you may need to run an analysis and design course. Specifically, this section contains the following:

data dictionary entry form
program overview form (first page)
program overview form (subsequent page)
screen layout form
print chart form

Needless to say, you have our permission to use these forms any way you see fit.

Data structure:

Group	Repetitions	Components	Comments

Program:	Page:
Designer:	Date:

Input/output specifications

File	Description	Use

Process specifications

Program:	Page:
Designer:	Date:

Process specifications

Screen name ______________________ Date ____________

Program name ____________________ Designer __________

	1	2	3	4	5	6	7	8	9	10	11	12	13	14	15	16	17	18	19	20	21	22	23	24	25	26	27	28	29	30	31	32	33	34	35	36	37	38	39	40	41	42	43	44	45	46	47	48	49	50	51	52	53	54	55	56	57	58	59	60	61	62	63	64	65	66	67	68	69	70	71	72	73	74	75	76	77	78	79	80
1																																																																																	
2																																																																																	
3																																																																																	
4																																																																																	
5																																																																																	
6																																																																																	
7																																																																																	
8																																																																																	
9																																																																																	
10																																																																																	
11																																																																																	
12																																																																																	
13																																																																																	
14																																																																																	
15																																																																																	
16																																																																																	
17																																																																																	
18																																																																																	
19																																																																																	
20																																																																																	
21																																																																																	
22																																																																																	
23																																																																																	
24																																																																																	

Fold back at dotted line.

Fold in at dotted line.

Document name ______________________ Date ______________

Program name ______________________ Designer ______________

Record Name	
	1
	2
	3
	4
	5
	6
	7
	8
	9
	10
	11
	12
	13
	14
	15
	16
	17
	18
	19
	20
	21
	22
	23
	24
	25
	26
	27
	28
	29
	30
	31
	32
	33
	34
	35
	36
	37
	38
	39
	40
	41
	42
	43
	44
	45
	46
	47
	48
	49
	50

Column positions: 1 2 3 4 5 6 7 8 9 10 11 12 13 14 15 16 17 18 19 20 21 22 23 24 25 26 27 28 29 30 31 32 33 34 35 36 37 38 39 40 41 42 43 44 45 46 47 48 49 50 51 52 53 54 55 56 57 58 59 60 61 62 63 64 65 66 67 68 69 70 71 72 73 74 75 76 77 78 79 80 81 82 83 84 85 86 87 88 89 90 91 92 93 94 95 96 97 98 99 100 101 102 103 104 105 106 107 108 109 110 111 112 113 114 115 116 117 118 119 120 121 122 123 124 125 126 127 128 129 130 131 132

1 2 3 4 5 6 7 8 9 10 11 12 13 14 15 16 17 18 19 20 21 22 23 24 25 26 27 28 29 30 31 32 33 34 35 36 37 38 39 40 41 42 43 44 45 46 47 48 49 50 51 52 53 54 55 56 57 58 59 60 61 62 63 64 65 66 67 68 69 70 71 72 73 74 75 76 77 78 79 80 81 82 83 84 85 86 87 88 89 90 91 92 93 94 95 96 97 98 99 100 101 102 103 104 105 106 107 108 109 110 111 112 113 114 115 116 117 118 119 120 121 122 123 124 125 126 127 128 129 130 131 132

Fold back at dotted line.

Fold in at dotted line.